A Multi-Physics and Continuum Mechanics Approach of Lithiated Silicon Nanowires

Donald C. Boone

Boone, Donald C. 1963-

A Multi-Physics and Continuum Mechanics Approach of Lithiated Silicon Nanowires

ISBN: 978-1495422799

TX0007882728 / 2014-02-02

Author's email address: db2585@caa.columbia.edu

A Multi-Physics and Continuum Mechanics Approach of

Lithiated Silicon Nanowires

Introduction

This study considers the electromagnetic stresses and simulates the lithium insertion into a silicon nanowire. The resulting model uses magnetohydrodynamic theory to explain the two detrimental effects that could result during the lithiated silicon process: (1) The partial lithiation effects that are observed in some silicon nanowires under negligible volume expansion; (2) The excessive volume expansion that is observed after full lithium ion insertion with a resulting Cassini oval shaped silicon nanowire. Magnetic and electric fields are introduced into this simulation via the electromagnetic term in order to introduce additional compressive stresses that slows down the lithiation process and results in a partially lithiated silicon nanowire under certain boundary conditions. Also, additional tensile stresses are introduced via magnetic dipole moments into this simulation to explain the anisotropic volume expansion that can occur under certain situations.

Silicon has a great future potential as a long-lasting anode material for automotive and commercial Li-ion batteries. Specific capacity for silicon (4200 mA/h/g) [6] is more than 10 times greater than the current commercialized carbon-based anode material (~372mA/h/g [1]). In recent years it has been reported during first lithiation of silicon nanoparticles and nanowires a great volumetric expansion in the order of 300% to 400% of their original volume [1,3,6]. This has led to poor cycling performance, mechanical fractures and ultimate failure during service. There have also been reports while restraining the volume expansion, partial lithiation of the silicon nanoparticle or nanowire [2]. This limits the potential of the silicon anode material, restricting it to approximately half of its specific charge capacity.

The current body of knowledge in research for lithiated silicon anode materials has been exclusively focused on lithium ion diffusion process. This report will present a possible alternative solution to both volume expansion and partial lithiation problems.

1. The Model

The computational models that will be presented are based on the elastic-diffusion model of a silicon spherical nanoparticle that was presented in Golmon et-al.[1]. This model described the lithiation process from an electrochemical point of view. Lithium ions diffused through the

crystalline silicon (c-Si) lattice, breaking silicon-silicon bonds and reforming lithium-silicon bonds, creating tensile stress throughout the nanoparticle while simultaneously Li-Si alloy material concentration builds up a compressive stress uniformly within the nanoparticle shell. Although the elastic-diffusion model does explain the volume expansion, it fails to explain for the slowing of the interfacial reaction front which is the interface between the c-Si and Li_xSi alloy that causes partial lithiation within the Si nanowire as was reported by Liu et-al. [2]. Therefore, an extension of this model is performed on a cylindrical silicon nanowire with the addition of electromagnetic theory. In the computational model or simulation that will be presented, a series of partial differential equations (PDE) and trail functional (TF) are utilized to determine the effects of four solutions that comprises the varies stresses within our silicon nanowire model. The solutions of PDE with the use of TF are: 1) Elastic Displacement **u**, 2) Concentration $\overline{c}$, 3) Magnetic Field **B,** and 4) Electric Field **E.**

Elastic Displacement (u) Differential and Trial Functional Equations

Time Dependent Volumetric Equation:

$$\frac{\partial u}{\partial t} = \varepsilon_{rr} + \varepsilon_{\theta\theta} + \varepsilon_{zz} \tag{1}$$

$$\varepsilon_{rr} = \frac{\partial u_r}{\partial r} + \frac{1}{2}\left[\left(\frac{\partial u_r}{\partial r}\right)^2 + \left(\frac{\partial u_\theta}{\partial r}\right)^2 + \left(\frac{\partial u_z}{\partial r}\right)^2\right] \tag{2}$$

$$\varepsilon_{\theta\theta} = \left(\frac{1}{r}\frac{\partial u_\theta}{\partial \theta} + \frac{u_r}{r}\right) + \frac{1}{2}\left[\left(\frac{1}{r}\frac{\partial u_r}{\partial \theta} - \frac{u_\theta}{r}\right)^2 + \left(\frac{1}{r}\frac{\partial u_\theta}{\partial \theta} + \frac{u_r}{r}\right)^2 + \left(\frac{1}{r}\frac{\partial u_r}{\partial \theta}\right)^2\right] \tag{3}$$

$$\varepsilon_{zz} = \frac{\partial u_z}{\partial z} + \frac{1}{2}\left[\left(\frac{\partial u_r}{\partial z}\right)^2 + \left(\frac{\partial u_\theta}{\partial z}\right)^2 + \left(\frac{\partial u_z}{\partial z}\right)^2\right] \tag{4}$$

$$\textbf{where} \quad u_r = \frac{2}{3}u \ , \quad u_\theta = \frac{2}{3}u \ , \quad u_z = \frac{u}{3} \tag{5}$$

Trial Functional u:

$$u(r, \theta, z, t, p) = \frac{C_u}{p}\left\{[r^4 + a^4 - 2a^2r^2(1 + \cos 2\theta) - b^4]e^{\left(\frac{z}{L} + \lambda_1 t\right)} + C_1\right\} \tag{6}$$

The equation within the square bracket [ ] and highlighted in red is the Cassini Oval equation.

Concentration $\bar{c}$ Differential and Trial Functional Equations

Fick's Second Law of Diffusion:

$$\frac{\partial \bar{c}}{\partial t} = D\nabla^2 \bar{c} \tag{7a}$$

$$\frac{\partial \bar{c}}{\partial t} = D\left(\frac{\partial^2 \bar{c}}{\partial r^2} + \frac{1}{r}\frac{\partial \bar{c}}{\partial r} + \frac{1}{r^2}\frac{\partial^2 \bar{c}}{\partial \theta^2} + \frac{\partial^2 \bar{c}}{\partial z^2}\right) \tag{7b}$$

Trial Functional $\bar{c}$:

$$\bar{c}(r, \theta, z, t) = \frac{4\bar{c}_{max}}{\pi} \sum_{n=0}^{\infty} \frac{\sin(2n+1)}{2n+1}[1 + \cot(p\pi) - \csc(p\pi)]\left[\cos\left(\frac{r}{R}\theta\right)\cos\left(\frac{z}{L}\right)\right]e^{\lambda_2 t} + C_2 \tag{8}$$

Electromagnetic (B and E) Differential and Trial Functional Equations

Magnetic Field (B) Wave Equation:

$$\frac{\partial^2 B}{\partial t^2} = c^2\nabla^2 B \tag{9a}$$

$$\frac{\partial^2 B}{\partial t^2} = c^2\left(\frac{\partial^2 B}{\partial r^2} + \frac{1}{r}\frac{\partial B}{\partial r} + \frac{1}{r^2}\frac{\partial^2 B}{\partial \theta^2} + \frac{\partial^2 B}{\partial z^2}\right) \tag{9b}$$

Trial Functional B:

$$B(r, \theta, z, t) = 2C_m[1 + \cot(p\pi) - \csc(p\pi)]\left[\cos\left(\frac{r}{R}\theta\right)\cos\left(\frac{z}{L}\right)\cos\left(\sqrt{3}\frac{ct}{L}\right)\right]e^{\lambda_2 t} + C_3 \tag{10}$$

where

$$C_m(r, I) = \frac{\mu_0\, I\, r}{2\,\pi\, R^2} \tag{11}$$

Electric Field (E) Wave Equation:

$$\frac{\partial^2 E}{\partial t^2} = c^2 \nabla^2 E \tag{12}$$

$$\frac{\partial^2 E}{\partial t^2} = c^2 \left(\frac{\partial^2 E}{\partial r^2} + \frac{1}{r} \frac{\partial E}{\partial r} + \frac{1}{r^2} \frac{\partial^2 E}{\partial \theta^2} + \frac{\partial^2 E}{\partial z^2} \right) \tag{13}$$

Trial Functional E:

$$E(r, \theta, z, t) = 2C_e[1 + \cot(p\pi) - \csc(p\pi)] \left[\cos\left(\frac{r}{R}\theta\right) \cos\left(\frac{z}{L}\right) \cos\left(\sqrt{3}\frac{ct}{L}\right) \right] e^{\lambda_2 t} + C_3 \tag{14}$$

where

$$C_e(r, q) = \frac{q}{4\pi\epsilon_0 r^2} \tag{15}$$

MHD Magnetic Stress Equation

$$E \times B = \frac{\rho}{\mu_0} \left[-\vec{\nabla}\left(\frac{B^2}{2}\right) + B^2 \frac{\vec{R}}{R_c} \right] \tag{16}$$

where

$$R_c = a \left[\cos(2\theta) + \sqrt{\left(\frac{b}{a}\right)^4 - \sin^2(2\theta)} \right]^{\frac{1}{2}} \tag{17}$$

is the Cassini Oval equation express as a solution for it radius R.

Radial Stress Equation σ_r

$$\sigma_r = \bar{E}\left[\frac{v}{2}\left(\frac{u}{R_c}\right) - \frac{\Omega}{6}\bar{c}\left(1 + v\right) + \frac{aB}{\rho P}\left(1 - 2v\right)\right] \tag{18}$$

where $P = \epsilon_0 \chi E$ (19)

Tangential Stress Equation σ_t

$$\sigma_t = \bar{E}\left[\frac{3v}{5}\left(\frac{u}{R_c}\right) - \frac{\Omega}{6}\bar{c}\left(1 + v\right) + \frac{aB}{\rho \epsilon_0 \chi E}\left(1 - 2v\right)\right] \tag{20}$$

The elasticity equations 1 thru 6 describes the displacement (u) of the silicon atoms due to the diffusion of lithium atoms as they pass through the c-Si lattice. Equation 1 is the time dependent volumetric strain equation that describes the geometric shape of the nanowire. These equations also satisfies compatibility conditions for model. The trial functional (u) used to solve this set of differential equations is derived from Cassini oval equation. The coefficients C_u and C_1 are functions of r, θ, z, t and p where p is a parameter present in all of the PDE in this computational model. A time frequency coefficient λ_1 used in the time evolution of the model.

Equations 7a and 7b are Fick's second law of diffusion that models the lithium atom diffusion activity. The trial functional (c) that becomes the solution to the diffusion PDE describes a uniform concentration during steady state at the conclusion of the diffusion process. A Fourier series replaces the Bessel Function J_0 as part of the trial functional (c) in order to describe a constant compressive stress, which is the reason for the progressive slow down and ultimate cessation of the lithiation process. A time frequency coefficient λ_2 used in the time evolution of the model.

Equations 9 thru 17 is the set of equations that incorporates the electromagnetic (EM) theory into model. Most of the research literature at the time of this publication has reported exclusively on diffusion simulated models with the exception of such research findings reported by Wang et-al. [8]. The EM theory is added to the elastic-diffusion theory of this model via the magnetohydrodynamic (MHD) theory, which is a macroscopic plasma physic theory. The wave equation general solutions for both the magnetic **B** and electric **E** fields are adopted as the trial

functionals and are used in the MHD magnetic stress equation to obtain specific solution to the EM field. Once again the Cassini oval equation is invoked as with the elasticity equations to describe the geometric shape of the field.

The radial and tangential stress equations displayed in equations 18 thru 20 have three components each: elasticity stress σ_e, diffusion stress σ_d and electromagnetic stress σ_{em}

$$\sigma_e = \bar{\mathbf{E}}\left[\frac{v}{2}\left(\frac{u}{R_c}\right)\right] \qquad \text{(radial stress)} \qquad (21a)$$

$$\sigma_e = \bar{\mathbf{E}}\left[\frac{3v}{5}\left(\frac{u}{R_c}\right)\right] \qquad \text{(tangential stress)} \qquad (21b)$$

$$\sigma_d = -\bar{\mathbf{E}}\left[\frac{\Omega}{6}\bar{c}\left(1+v\right)\right] \qquad (22)$$

$$\sigma_{em} = \bar{\mathbf{E}}\left[\frac{a\mathbf{B}}{\rho\epsilon_0\chi\mathbf{E}}\left(1-2v\right)\right] \qquad (23)$$

As alluded to previously, the radial σ_e^r and tangential σ_e^t elastic stresses are due to the strain that is developed from the displacement $\mathbf{u}$ from the silicon atoms while the diffusion stress σ_d is due to the lithium atom concentration that diffuses within the silicon lattice structure. The electromagnetic stress σ_{em} is derived from the current or electron flow along the silicon nanowire and the electric polarization of the silicon and lithium atoms that effectively creates electric dipole moments of these constituent. These stress components are of a compressive and tensile nature. In this report an examination of two computational models will be performed with regards to stress. The first, a negligible volume contraction model which due to a majority of compressive stresses and the second model volume expansion which are caused by overwhelming tensile stresses, within the silicon nanowire.

Constitutive parameter are present in the elastic, diffusion and electromagnetic sets of equations. Resistivity ρ, Poisson ratio v, modulus of elasticity $\bar{\text{E}}$, electric susceptibility χ ,diffusion coefficient D, etc., are material dependent which can be changed in this computational model to explore other material combinations such as replacing lithium for sodium which is popular research topic within the battery research community.

2. Magnetization and Magnetic Moments

Defining the radial σ_r and tangential σ_t stresses in equations 18 and 20, the focus is now brought upon the last term in both these equations, the electromagnetic term $\frac{a}{\rho P}\text{B}(1\text{-}2v)$, which is the main subject of this report. The product of this term with the modulus of elasticity $\bar{E}$ produces the electromagnetic stress σ_{em}. If we reposition the square brackets (disregarding the Poisson ratio term) in equation 24a an important insight in brought forth

$$\sigma_{em} = \bar{\text{E}}\left[\frac{a}{\rho P}\text{B}\right] \longrightarrow \sigma_{em} = \left[\bar{\text{E}}\frac{a}{\rho P}\right]\text{B} \qquad (24)$$

(a)   (b)

The parameter in the square bracket for equation 24b is magnetization [10]. Magnetization M is defined as magnetic moments [9] per unit volume, $\text{M} = \frac{m}{V}$. In this hypothesis, the magnetization has two components that represent the electromagnetic tension stress σ_{em}^t and electromagnetic compression stress σ_{em}^c that will be discussed in the next section of this report. The magnetization that is responsible for the σ_{em}^t is define within classical electromagnetic theory using the magnetic dipole moment $\mathbf{m}$ as the product of electric current loop $\mathbf{i}$ and the area $\mathbf{A}$ enclosed by the loop, $\mathbf{m} = \text{Ai}$ (figure 1a). The magnetization that describes the electromagnetic compression stress σ_{em}^c is define as the cross product of the current density $\mathbf{J}$ and the position vector $\mathbf{r}$ of the current density $\mathbf{M}_c = \sum(\mathbf{r} \times \mathbf{J})$ (figure 1b). The total magnetization $\mathbf{M}$ is define as

$$\mathbf{M} = \left[\bar{\text{E}}\frac{a}{\rho P}\right] = \sum_n^N \left(\frac{\text{Ai}}{V}\right)_n + \sum_n^N (\mathbf{r_n} \times \mathbf{J_n}) \qquad (25)$$

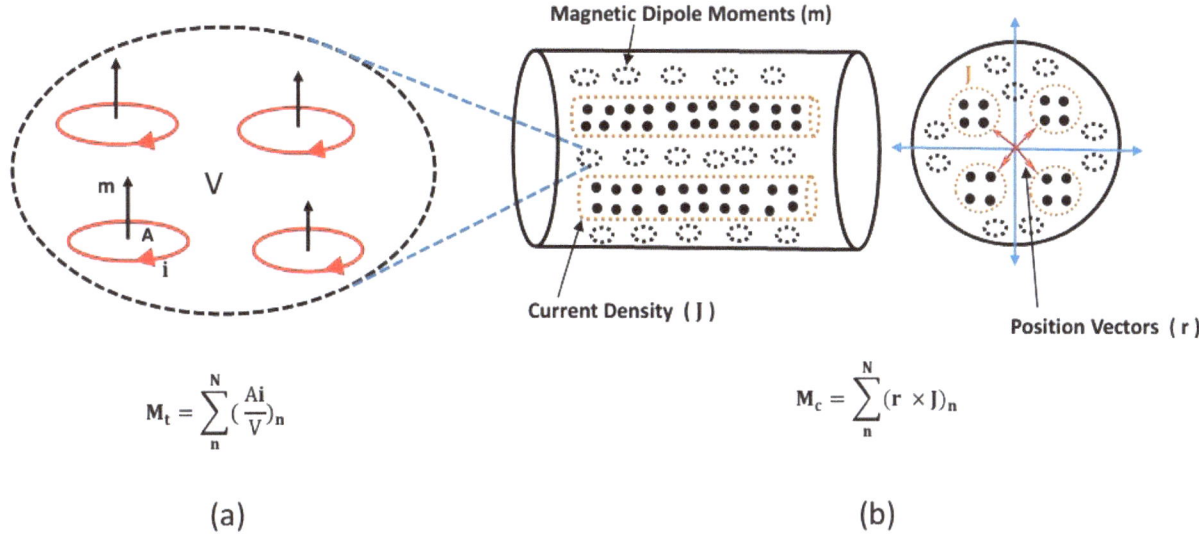

$$M_t = \sum_{n}^{N} \left(\frac{Ai}{V}\right)_n$$

(a)

$$M_c = \sum_{n}^{N} (r \times J)_n$$

(b)

Figure 1. (a) The magnetic dipole moments travel in cluster groups that define the magnetization M_t that is responsible for the electromagnetic tension stress σ_{em}^{t}. **(b)** The electron flow as current density **J** defines the magnetization M_c that induces the electromagnetic compression stress σ_{em}^{c}.

The magnetic dipole moments (or magnetic moments) **m** is clustered in separate nanoscale volumes and are traveling parallel to each other so the individual magnetic moments do not cancel out within the their group.

3. Magnetohydrodynamics

With the introduction of magnetization and magnetic moments, this brings us to the plasma physic theory that will be used in our computational model and was mentioned earlier in our report; Magnetohydrodynamics (MHD) Theory. The actual experimental apparatus for silicon nanowire is defined according to the in-situ transmission electron microscope (TEM) experimental setup (Figure 2a) that is similar to that in Liu et-al.[3] In the setup, the Si nanowire is in direct contact with Li metal where Li ions can enter the nanowire under an applied electric field.

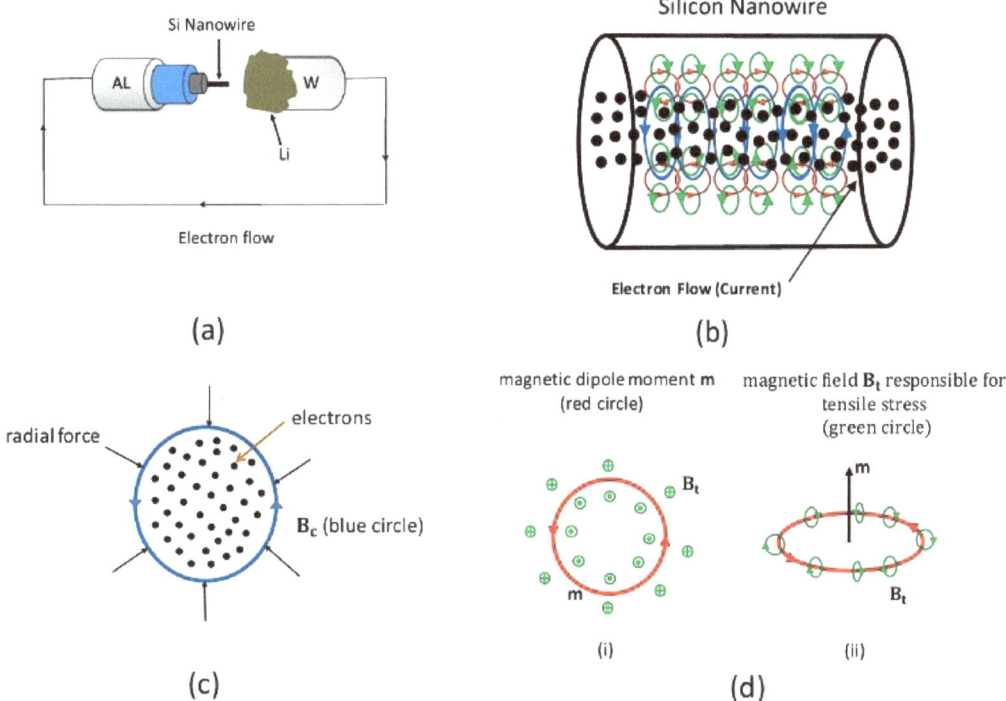

Figure 2. (a) In-situ experimental arrangement for a solid electrochemical cell using lithium metal counter electrode **(b)** Illustration of silicon nanowire during lithiation process. Electron flow induces magnetic field $\mathbf{B_c}$ (blue circles, see figure 2c) which in turn induces magnetic dipole moments $\mathbf{m}$ (red circles, see figure 2d). The magnetic dipole moments $\mathbf{m}$ induces the magnetic field $\mathbf{B_t}$ (green circles) that opposes and partially eliminates $\mathbf{B_c}$. **(c)** Si NW cross sectional view of electron flow (black dots). The current according to Lenz's law induces the magnetic field $\mathbf{B_c}$ (blue circle) which generates the electromagnetic compressive stress σ_{em}^c in the radial direction. The radial force is called the pinch effect. **(d)** (i) Cross sectional and top view of a single magnetic dipole moment $\mathbf{m}$. (ii) Displays the resulting magnetic field $\mathbf{B_t}$ induced by the magnetic dipole moment $\mathbf{m}$. $\mathbf{B_t}$ is responsible for the electromagnetic tension stress σ_{em}^t.

Prior to the beginning of lithiation process, the concentration of Li_xSi is zero (x=0). Once the lithiated process begins, lithium ions starts to diffuse within the silicon nanowire. The diffusion process also begins the chemical reaction between lithium and silicon, breaking Si-Si bonds and forming Li-Si bonds, (Li_xSi where x>0) [4]. As describe by MHD the electromagnetic stress σ_{em} is composed of two components, the electromagnetic tension stress σ_{em}^t and electromagnetic compression stress σ_{em}^c . Since the electromagnetic stress σ_{em} is a function of the magnetic field **B**, equation 26 is composed of two opposing terms, compression and tension components defined as $\mathbf{B_C}$ and $\mathbf{B_t}$, respectively

$$\mathbf{B} = \overleftrightarrow{\mathbf{B_c}}(\mathbf{J}, \mathbf{r}) + \overleftrightarrow{\mathbf{B_t}}(\mathbf{m}, V, n) \qquad\qquad (26)$$

As illustrated in figure 2b and 2c, the magnetic field $\mathbf{B_c}$ represented by large blue circles is responsible for the electromagnetic compressive stress σ_{em}^c and is a function of the current density $\mathbf{J}$ and the position vector $\mathbf{r}$ of the current density. The magnetic field $\mathbf{B_t}$ depicted as green circles are responsible for the electromagnetic tension stress σ_{em}^t and is dependent upon the individual nanosize current loops $\mathbf{i}$ that compose the magnetic moments $\mathbf{m}$ as were depict in figure 2b and 2d. In electromagnetic theory, magnetic moments $\mathbf{m}$ are swirling electron loops that experience Lorentz's force and are analogous to turbulence in fluid dynamics or eddy currents in electrical circuit theory. In the model that is presented, these currents will be induced when the nanowire experiences change in the magnetic field or a decrease in the resistivity ρ. According to electromagnetic theory, a decrease in resistivity ρ (or an increase in conductivity) acts to strengthen the current density $\mathbf{J}$. Lenz's law states that the magnetic moment current loop will circulate in such a way as to create an induced magnetic field $\mathbf{B_t}$ that opposes the magnetic field $\mathbf{B_c}$. Magnetic moments $\mathbf{m}$ will be generated wherever the nanowire experiences a change in the magnitude or direction of the magnetic field $\mathbf{B}$ at any location within the nanowire, not just at the surface as depicted in figure 2b. $\mathbf{B_t}$ is the function of the individual magnetic moments $\mathbf{m}$, nanocluster volumes V of $\mathbf{m}$, and the number $\mathbf{n}$ of magnetic moments within the nanowire.

In using MHD theory, the electromagnetic compressive stress σ_{em}^c is derived from electrons traveling through the nanowire, as displayed in figure 2b, are analogous to parallel electrical wires which induces magnetic field $\mathbf{B_C}$ (blue circle) and due to the Lorentz's force cause them to be attracted to each other. A collection of electrons simulating parallel electric wires would therefore attract each other resulting in a compressive stress that acts radially to reduce the diameter of the nanowire (Figure 2c). In MHD theory this phenomena is known as the pinch effect [8].

Similarly as with the electromagnetic compressive stress σ_{em}^c, utilizing MHD theory, the electromagnetic tension stresses σ_{em}^t is induced by magnetic moments $\mathbf{m}$. When electrons of this magnetic moment $\mathbf{m}$ travel in circular path or loop as discussed previously, they induce a magnetic field $\mathbf{B_t}$ (figure 2d) that is in opposition to $\mathbf{B_c}$. The magnetic field inside the magnetic moments loop is stronger than outside the loop. This in turn creates an outward tangential or hoop stress which will be the electromagnetic tensile stress σ_{em}^t. In addition, the electrons on opposite sides of the loop travels in opposing directions or antiparallel and create a repulsive force between the electrons on the other side of the loop which also adds to the electromagnetic tension stress σ_{em}^t.

With defining of the electromagnetic stress σ_{em}, this report continues with the constitutive nature of the model. There are only two materials in our simulation, silicon and lithium. The major material property that is the defining parameter is resistivity δ. The resistivity of silicon, a semiconductor material, is 10^2 Ω-m whereas for lithium, a metal and an excellent conductor, the resistivity is 10^{-8} Ω-m. This is a 10 billion fold difference in resistivity δ and the reason for the great electromagnetic stress σ_{em} that is generated from an infinitesimal magnetic $\mathbf{B}$ and electric field $\mathbf{E}$. At the start of the lithiation process, the electrons travel with an initial potential difference of ~2V [3]. Electrons first encounter c-Si with a relatively high resistivity ρ which creates a very low electromagnetic stress σ_{em} of 10^{-3} Pa. Soon after lithium atoms start to diffuse through the silicon lattice and electrons flows through lithium with an extremely low resistivity, the electromagnetic stress σ_{em} increases dramatically.

The lithiation of the silicon nanowire is a diffusion driven process. During the computational analysis of any set of boundary conditions, it was discovered that elasticity σ_e and diffusion σ_d stresses alone does not account for all of the volumetric strain. There are four simulations that are applied to silicon nanowire model based on elastic-diffusion-electromagnetic process: (1) radial stress of a partially lithiated Si-NW with no volume expansion, (2) tangential stress of a partially lithiated Si-NW with no volume expansion, (3) radial stress of a fully lithiated Si-NW with over 300% volume expansion, (4) tangential stress of a fully lithiated Si-NW with over 300% volume expansion. The simulations for this report are a view of our model at times when the lithium insertion is at maximum concentration of Li_xSi at x=3.75 and the lithium diffusion process is near its completion.

4. Partially Lithiated Silicon Nanowire

The simulation of the partially lithiated silicon nanowire models the balance of tensile and compressive forces that results in a near zero volume change in the nanowire. As a result of negligible volume change (the volume strain is approximately zero) the advancement of the interfacial front between the c-Si and lithiated silicon slows to almost a complete stop according to Liu et.al. [2]. This is due to the cessation of the chemical reaction between silicon and lithium where the production of lithiated silicon all but comes to an end. Figure 3 shows the radial and tangential stresses respectively that develops at the end of lithiation process. However at the beginning of the lithiation process, lithium ions diffuses from Li cathode material and crosses the electrolyte and are inserted at the point of entry of the Si nanowire. The electrons that travels through the experimental circuit and flows through the aluminum rod and silicon nanowire as current complete their journey when they reunites with the lithium ions to form lithium atoms. Li starts to diffuse at the surface in a longitudinal z-direction and as lithium insertion continuesthe Li atoms also diffuse in a radial r-direction. The diffusion rate in the z-direction is faster than the

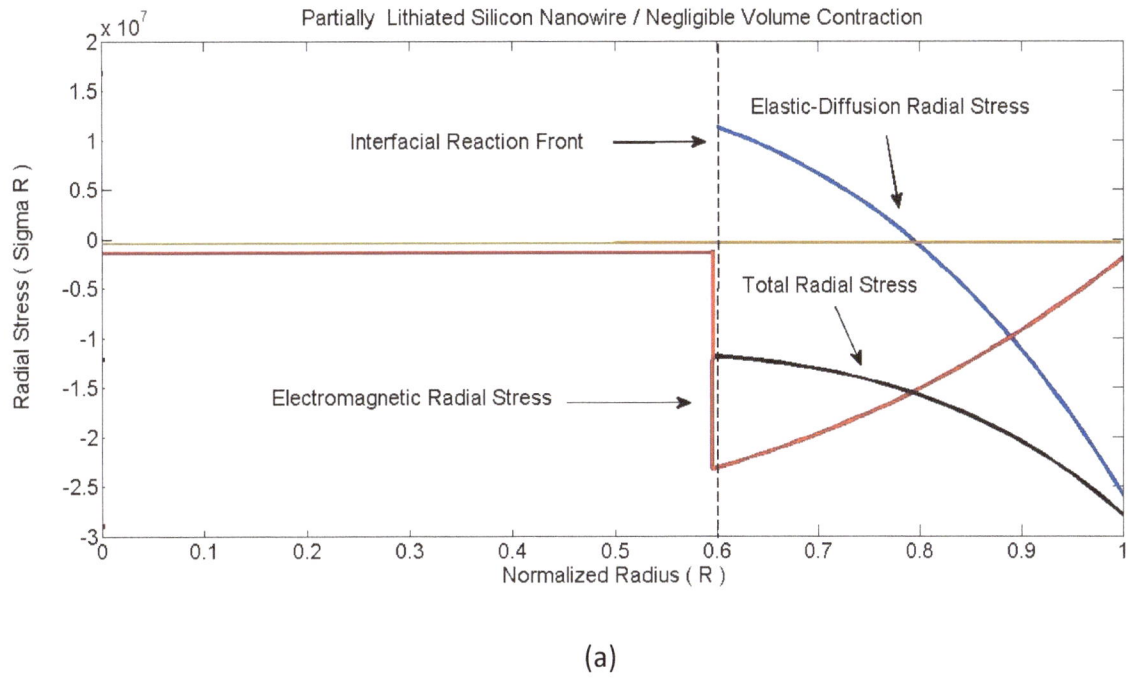

(a)

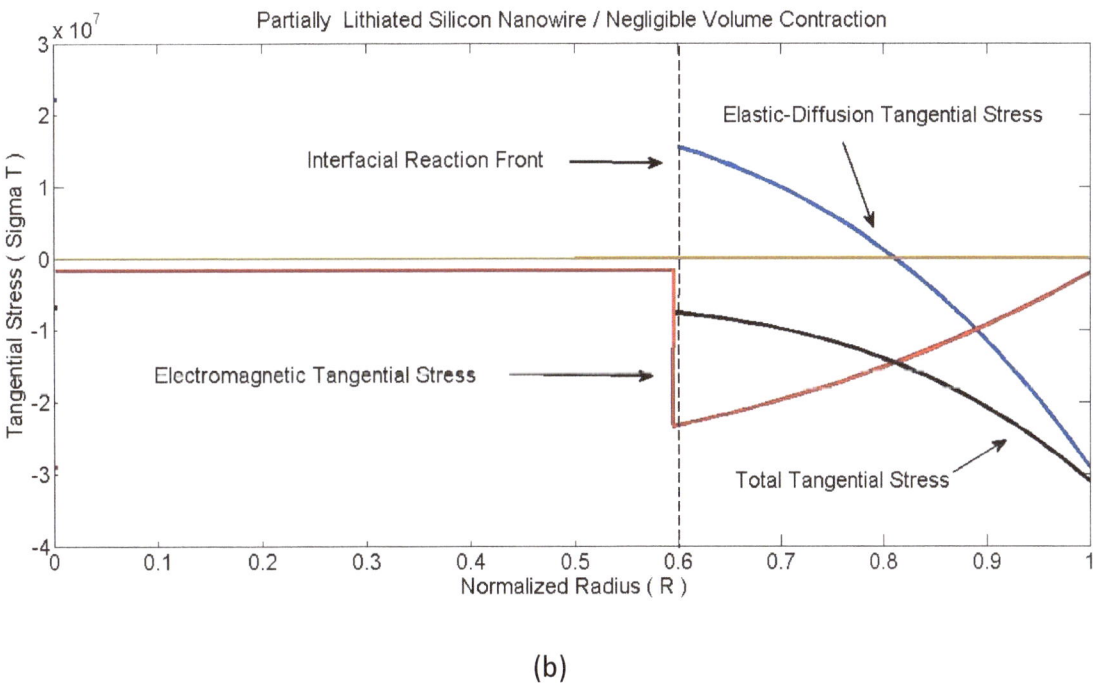

(b)

Figure 3. (a) Total radial stress for partially lithiated silicon nanowire. The total radial stress is near zero which results in negligible volume contraction. (b) However the total tangential stress has a non-zero value and is under compressive stress which could create cracks on the surface. Location of radial and tangential stresses are at $(r, \theta, z) = (r, 0, 1)$

diffusion rate in the radial r-direction [3]. The electron flow is analogous to current in an electric circuit as electrons travel in the direction of least resistance and flow through the low resistant lithium metal instead of the higher resistant silicon semiconductor material. As Li atoms diffuses to the center at a slower rate than the Li atoms at the surface, the electrons will follow the Li metal and electric current will be establish in the Li_xSi shell of the nanowire. Radial and tangential stresses are in units of pascals (Pa).

The stress due to concentration $\bar{c}$ of Li_xSi which is called the diffusion radial and tangential stresses σ_d, are constant at the end of the lithiation process. This is because the production of lithiated silicon has reached its maximum which is define to be x=3.75 which agrees with most of the research literature. The diffusion stress σ_d in this simulation is compressive which tends to slow down the lithiation process. The elastic radial and elastic tangential stresses σ_e are at certain location within the nanowire tensile and at other locations compressive. While compressive stress slows down the lithiation process, tensile stresses promotes the lithiation process. The electromagnetic stress σ_{em} is at a maximum compressive stress at the interfacial front and gradually decrease to their minimum value at the surface. The elastic-diffusion radial and tangential stresses, which is a sum of $\sigma_e + \sigma_d$, start as a tensile stress at the interfacial reaction front and gradually decrease until it reaches zero and converts to an increasing compressive stress underneath the surface at approximately 0.8 of the normalized radius. In order to have near zero volume expansion in the radial stress simulation, summation of the elastic, diffusion and electromagnetic radial stress components must come close to zero. However, the summation of the tangential stress components is non-zero and compressive.

5. Volume Expansion of Lithiated Silicon Nanowire

We continue are computational modeling with the simulation of a completely lithiated silicon nanowire. Unlike the partially lithiated nanowire, the interfacial reaction front in the fully lithiated silicon nanowire from the beginning of lithium insertion at the surface advances to the center of the nanowire until completion of lithiation. As a result of full lithiation, as reported by varies research groups that studied nanoparticles and nanowires, there is a great expansion in volume. For spherical nanoparticles there is a symmetrical and uniform volume expansion however for silicon nanowires the expansion is anisotropic that result in a Cassini oval shape that resemble the geometric curves discovered by astronomer Giovanni Domenico Cassini (June 8, 1625 – September 14, 1712) in the 17th century (figure 4).

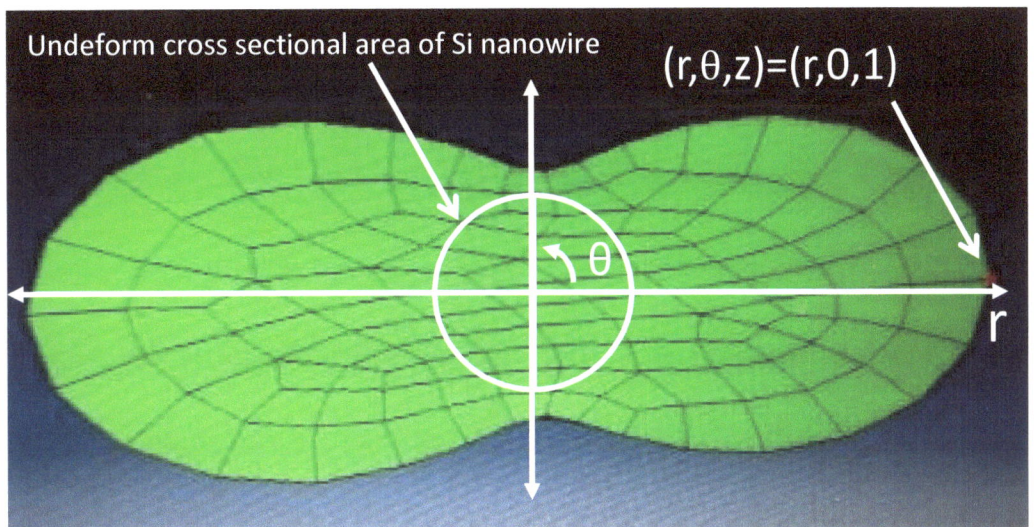

Figure 4. Abaqus model of Cassini ovals shaped silicon nanowire after the 300% expansion. $(r,\theta,z) = (r,0,1)$ is the location of the maximum radial and tangential stress in cylindrical coordinates.

The anisotropic expansion has been reported to be over 300% of the original volume by most research studies. The diffusion radial and tangential stresses σ_d are constant and compressive however with different magnitude as that with the partially lithiated Si nanowire. The elastic radial and tangential stresses σ_e are tensile and of a greater magnitude then the partially lithiated nanowire that has near zero volume contraction. The elastic stress component σ_e alone accounts for over 300% of the volume expansion. However the diffusion stress component σ_d since it is compressive tend to contract or reduce the volume of the nanowire. The summation of the elastic and diffuse components $\sigma_e + \sigma_d$ results in a 180% volume expansion. The electromagnetic term in the radial and tangential stress equations supplies the additional strain of approximately 120% that is require to justify a 300% volume expansion (figure 5) as reported by so many research groups. The electromagnetic stress, as with our previous simulation of a Si nanowire with near zero volume expansion, is also comprised of a compressive and tensile components. However with this simulation the electromagnetic tension stress σ_{em}^t component predominate over the electromagnetic compressive stress σ_{em}^c component from the center to the surface of the nanowire. The volumetric expansion is not constant from the center to surface of the nanowire. The volume expansion is measure at 300% to 375% at varies locations along the radial distance. The stress does not terminate to zero at the surface of the nanowire, which leads to the conclusion that in order for the silicon nanowire to avoid fracture at the surface, de-lithiation of the nanowire is possibly necessary. Radial and tangential stresses are in units of pascals (Pa).

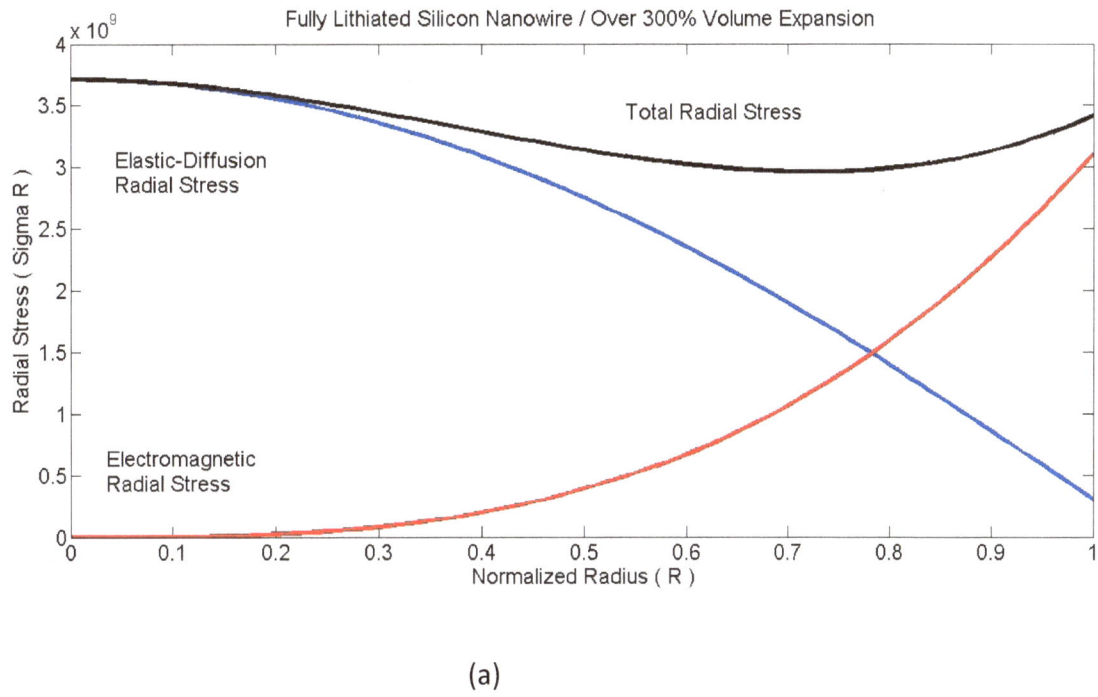

(a)

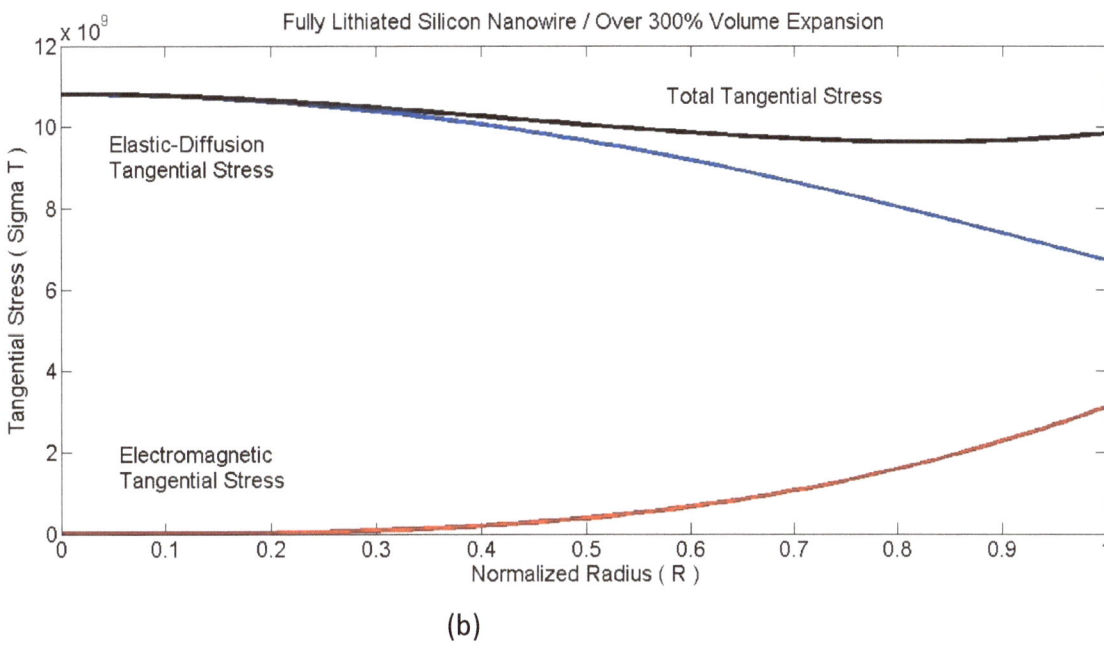

(b)

Figure 5. (a) The total radial stress for fully lithiated silicon nanowire (b) The total tangential stress for fully lithiated silicon nanowire. For (a) and (b) the total radial and tangential stress are calculated at their maximum which is located at $(r,\theta,z)=(r,0,1)$ (see figure 4) and it resulted in over 300% volume expansion.

6. Summary

In this report a case is established for the possible existence of an electromagnetic stress σ_{em} that is derived from fundamental plasma and electromagnetic theory. The essential concepts in this theory are that of MHD and magnetization. MHD theory in general is a plasma physic theory that explains the interaction of electrically charged particles under the influence of magnetic and electrical fields. However in this model the use of MHD is to describe the electrons that induce magnetic fields $\mathbf{B}$ that are used to generate the electromagnetic stresses σ_{em} in the radial and tangential directions in Si nanowire. The concept of magnetization $\mathbf{M}$ in conjunction with MHD gives the electromagnetic stress σ_{em} a tensile and compressive stress components. The flow of electrons within the silicon nanowire induces a magnetic field $\mathbf{B_c}$ that produces the compressive electromagnetic stress that aids in the control of volume expansion at the expense of full lithiation of the nanowire (partial lithiation). The magnetic field $\mathbf{B_c}$ in turn induces multiple nanosize electrical current loops called magnetic moments (or magnetic dipole moments) $\mathbf{m}$ that induce an opposing magnetic field $\mathbf{B_t}$ that acts in opposition to the magnetic field $\mathbf{B_c}$. The magnetic field $\mathbf{B_t}$ is responsible for the tensile electromagnetic stress that is complementary to the elastic tensile stress that is created by lithium diffusion process. The magnitude of the magnetic dipole moments $\mathbf{m}$ is a function of 1) the strength of the magnetic field $\mathbf{B_c}$, 2) the rate of change of the magnetic field $\mathbf{B_c}$ and 3) the value of the resistivity ρ. The greater the magnetization, the stronger the electromagnetic stress σ_{em}. It should be emphasize that the magnetic dipole moments $\mathbf{m}$ diminishes (via $\mathbf{B_t}$ that is in opposition to $\mathbf{B_c}$) the current density $\mathbf{J}$. Therefore as the magnetic moments $\mathbf{m}$ increase in strength, $\mathbf{B_t}$ also increase at the expense of $\mathbf{B_c}$ which decreases due to the reduced current density $\mathbf{J}$. The resistivity ρ is the primary constitutive parameter in our model. A low resistivity ρ or conversely a high conductivity of lithium is the reason for an electromagnetic stress σ_{em} in the order of 10^9, which is the same order for elastic σ_e and diffusion σ_d stresses. The magnetic field $\mathbf{B}$ that induces the electromagnetic stress is infinitesimally small at 10^{-16} tesla. It's this hypothesis that a minute magnetic field $\mathbf{B}$ that possibly cannot be detected through measuring instrumentation can explain the two detrimental effects of partially lithiation and volume expansion that until now has been attributed to exclusively diffusion process.

In this book the term "electromagnetic stress" is used to differentiate from magnetic stress. Although the magnetic field was used extensively to describe the process of lithiation, the electric interaction (or electric potential) between charges particles is the dominating interaction in this lithium diffusion process while magnetic interaction (or magnetic stress) is negligible. Therefore use of the term 'magnetic stress' would be a misleading terminology since the electric interaction is of a significant value much greater than that of magnetic stress. In order to have a comprehensive description of lithiated silicon process, a quantum mechanical approach using density functional theory (DFT) and molecular dynamics should be incorporated.

Acknowledgement

I would like to thank Dr. Reza Shahbazian-Yassar of Michigan Technological University - Mechanical Engineering Department for introducing me to the problems of lithiated silicon nanowires.

References

[1] Stephanie Golmon, Kurt Maute, Se-Hee Lee, Martin L. Dunn, Stress generation in silicon particles during lithium insertion, Applied Physics Letters 97 033111, 2010

[2] Xiao Hua Liu, Feifei Fan, Hui Yang, Sulin Zhang, Jian Yu Huang, Ting Zhu, Self-Limiting Lithiation in Silicon Nanowires, ACS NANO 7 (2) p1495–1503 2013

[3] Xiao Hua Liu, He Zheng, Li Zhong, Shan Huang, Khim Karki, Li Qiang Zhang, Yang Liu, Akihiro Kushima, Wen Tao Liang, Jiang Wei Wang, Jeong-Hyun Cho, Eric Epstein, Shadi A. Dayeh, S. Tom Picraux, Ting Zhu, Ju Li, O John P. Sullivan, John Cumings, Chunsheng Wang, Scott X. Mao, Zhi Zhen Ye, Sulin Zhang, Jian Yu Huang, Anisotropic Swelling and Fracture of Silicon Nanowires during Lithiation, Nano Letter 11 p3312-3318 2011

[4] Kejie Zhao, Matt Pharr, Qiang Wan, Wei L. Wang, Efthimios Kaxiras, Joost J. Vlassak, Zhigang Suo, Concurrent Reaction and Plasticity during Initial Lithiation of Crystalline Silicon in Lithium-Ion Batteries, Journal of The Electrochemical Society, 159 (3) p.A238-A243 2012

[5] Shan Huang, Ting Zhu, Atomistic mechanisms of lithium insertion in amorphous silicon, Journal of Power Sources 196 p3664–3668 2011

[6] Sung Chul Jung, Jang Wook Choi, Young-Kyu Han, Anisotropic Volume Expansion of Crystalline Silicon during Electrochemical Lithium Insertion: An Atomic Level Rationale, ACS Publication Nano Letter 12 p5342-5347 2012

[7] Zhiguo Wang, Meng Gu, Yungang Zhou, Xiaotao Zu, Justin G. Connell, Jie Xiao, Daniel Perea, Lincoln J. Lauhon, Junhyeok Bang, Shengbai Zhang, Chongmin Wang, Fei Gao, Electron-Rich Driven Electrochemical Solid-State Amorphization in Li–Si Alloys, Nano Letter 13, p4511–4516 2013

[8] Paul M. Bellan, Fundamentals of Plasma Physics, Cambridge University Press 2004, Print

[9] David J. Griffiths, Introduction to Quantum Mechanics 2nd edition, Prentice Hal 2005, Print

[10] John R. Hook, Henry E. Hall, Solid State Physics 2nd edition, John Wiley & Sons 1991, Print

[11] Michael Lai, David Rubin, Erhard Krempl, Introduction to Continuum Mechanics 4thedition, Elsevier 2010, Print

About the Author

Donald Boone received a Bachelor of Technology degree in Aeronautical Engineering Technology from New York Institute of Technology, a Master of Engineering degree in Mechanical Engineering from University of Maryland and a graduate Mechanical Engineer degree with a specialization in Micro/Nanoscale Engineering from Columbia University. He is also a license Professional Engineer in the Commonwealth of Virginia and the State of Maryland. He is currently working as a research engineer at the Nanoscience Research Institute in Columbia, Maryland.

Any comments about this publication should be email to db2585@caa.columbia.edu.

www.ingramcontent.com/pod-product-compliance
Lightning Source LLC
Chambersburg PA
CBHW040754200526
45159CB00025B/2088